AREAS WITH LIGHT

ZONAS CON LUZ

ESPACES LUMINEUX

BEREICHE MIT LICHT

4

AUTHORS
Fernando de Haro & Omar Fuentes

EDITORIAL DESIGN & PRODUCTION
AM Editores S.A. de C.V.

PROJECT MANAGERS
Valeria Degregorio Vega
Tzacil Cervantes Ortega

COORDINATION
Martha P. Guerrero Martel
Edali Nuñez Daniel

COPYWRITERS
Abraham Orozco
Roxana Villalobos

ENGLISH TRANSLATION
Louis Loizides

FRENCH TRANSLATION
Wordgate translations

GERMAN TRANSLATION
Heike Ruttkowski

EDITORES PUBLISHERS

100+ TIPS · IDEAS
areas with light . zonas con luz
espaces lumineux . bereiche mit licht

© 2009, Fernando de Haro & Omar Fuentes
AM Editores S.A. de C.V.
Paseo de Tamarindos 400 B, suite 102, Col. Bosques de las Lomas, C.P. 05120, México, D.F.
Tels. 52(55) 5258 0279, Fax. 52(55) 5258 0556. ame@ameditores.com **www.ameditores.com**

ISBN 13: 978-607-437-021-8

Printed in China.

INTRODUCTION
INTRODUCCIÓN
INTRODUCTION
EINLEITUNG

Daylight is the best form of illumination around. At the same time, it also changes and is not very easy to control during the course of the day and at different times of the year. Sunlight is not the same in spring and summer as in fall and winter, and its intensity will highlight different architectural features such as volumes, approach areas, facades and even indoor spaces. The solution is never as straightforward as it looks and a few tips on how to make the most of daylight will always come in handy. Artificial light can be modified to suit the look required for a given building. When projected onto porches, facades, patios, gardens and, especially, indoors, it can create the exact ambience that users want. In technical terms, there are many different types of lighting for both outdoors and indoors, such as incandescent, halogen, neon and energy-saving lights, to mention but a few. When combined with certain materials or reflective structures, glass or metal, they can create effects that complement the building's architectural concept. But this is no simple feat, which is why this volume provides ideas to help readers find the solution that best suits them.

La luz natural es el iluminante por excelencia, es el más perfecto con que se puede contar y sin embargo, es el más cambiante y menos controlable a lo largo del día y en las diferentes épocas del año. La luz del sol no es igual en la primavera y el verano que en el otoño y el invierno y de acuerdo con su intensidad hará que destaquen, más o menos, los diferentes elementos de una composición arquitectónica, como la volumetría, las áreas de aproximación, las fachadas e incluso los espacios interiores. La solución no es tan simple como parece y siempre es conveniente recibir algunos tips de la forma más conveniente de aprovechar la luminosidad natural. La luz artificial, en cambio, se puede modelar de acuerdo con el carácter que se pretenda dar al edificio, en los pórticos, las fachadas, en los patios y jardines, y sobre todo en los interiores donde se pueden crear ambientes de acuerdo con los gustos y las necesidades. Técnicamente existen diferentes tipos de luminarias tanto para los exteriores como los interiores, como las incandescentes, las de halógeno, las de neón, los focos ahorradores, entre una gran variedad que, combinadas con ciertos materiales o estructuras reflejantes, cristales o metales, crean efectos que incluso pueden llegar a complementar el concepto arquitectónico del edificio. No es tampoco una tarea sencilla y por eso este volumen ofrece algunas ideas que sin duda ayudarán a encontrar la solución más conveniente.

La lumière naturelle est par excellence l'illumination la plus parfaite sur laquelle on peut compter et pourtant, c'est aussi la plus changeante et la moins maîtrisable tout au long de la journée et des différentes époques de l'année. La lumière du soleil au printemps et en été diffère de celle de l'automne et de l'hiver et provoquera la mise en valeur, plus ou moins importante, selon son intensité, des éléments d'une composition architectonique, comme la volumétrie, les aires d'approximation, les façades et même les espaces intérieurs. La solution n'est pas aussi simple qu'elle y paraît et il est opportun de recevoir quelques astuces sur la manière la plus convenable de profiter de la lumière naturelle. La luminosité artificielle, quant à elle, peut se moduler selon la caractéristique que l'on veut donner au bâtiment, aux portiques, aux façades, aux patios et jardins, et surtout aux intérieurs dans lesquels on peut créer des ambiances en accord avec les goûts et besoins. D'un point de vue technique, il existe différents types de luminaires aussi bien pour l'extérieur que pour l'intérieur, comme les incandescents, les halogènes, les néons, les ampoules économie d'énergie, entre autres qui, combinés avec des matériaux spéciaux ou des structures réfléchissantes, vitres ou métaux, créent des effets qui peuvent même arriver à compléter le concept architectonique du bâtiment. Cette tâche n'étant pas forcément facile à réaliser, ce manuel propose quelques idées qui sans aucun doute aideront à trouver la solution la plus opportune.

Natürliches Licht ist die perfekteste Lichtquelle, die man sich vorstellen kann; dennoch handelt es sich auch im Verlauf des Tages und zu den verschiedenen Jahreszeiten um die wechselhafteste und am wenigsten kontrollierbare Beleuchtung. Das Sonnenlicht im Frühling unterscheidet sich von dem im Sommer, Herbst und Winter; und in Übereinstimmung mit der Intensität der Sonne, werden die unterschiedlichen Elemente einer architektonischen Komposition mehr oder weniger hervorgehoben, wie die einzelnen Ebenen, die Fassaden und die Innenbereiche. Die Lösung dieses Problems ist nicht so einfach wie es erscheint und daher sind ein paar Tipps immer nützlich. Das künstliche Licht kann hingegen so gestaltet werden, wie es zur Beleuchtung von Gebäuden, Fassaden, Innenhöfen und Gärten benötigt wird; dies ist vor allem auch im Innenbereich wichtig, wo die Atmosphäre in Übereinstimmung mit den Vorlieben und Bedürfnissen der Bewohner gestaltet werden kann. Technisch gesehen, existieren verschiedene Arten von Beleuchtung, sowohl für den Aussen- als auch für den Innenbereich, Glühlampen, Halogenlampen, Neonleuchten, Sparlampen usw.; kombiniert mit bestimmten Materialien oder reflektierenden Strukturen, Glas oder Metall, können Effekte geschaffen werden, die das architektonische Konzept des Gebäudes sogar vervollständigen können. Dies ist auch keine einfache Aufgabe und daher bietet dieser Band einige Ideen, die sicherlich dabei helfen werden, die günstigste Lösung ausfindig zu machen.

OUTDOORS

EXTERIORES

EXTÉRIEURS

AUSSENBEREICHE

Artificial light is a great outdoor option for enhancing the building's details, the color and texture of the materials used, the design of the façade and approach areas, the entrance, windows and other elements that afford balance and beauty for the building. Lighting can be provided by lights set at the bottom or upper sections of the building, from inside the house or from flying buttresses fixed to trees or any other type of container.

Daylight will brighten up the whole building and create a play of light and shade throughout the day, in accordance with the construction's specific details, as well as the color, texture and finishes of the materials, and the surrounding features.

Light and its behavior on outdoor volumes are usually regarded as part of the architectural design, and must therefore be defined in the building program. The chosen solution must specify the type of lighting and the shadows that will be thrown against walls in order to bring out the house's personality to the full.

En el exterior, la luz artificial ofrece la oportunidad de acentuar detalles del edificio, el color, la textura de los materiales, así como el diseño de la fachada y su zona de aproximación, la puerta de acceso, ventanales y otros elementos que dan armonía y belleza al conjunto. La luz puede provenir de luminarias puestas al pie o en lo alto de la propia edificación, del interior de la casa o de arbotantes colocados en los árboles o en cualquier otro elemento.

La luz natural ilumina todo el conjunto y crea un juego de luces y sombras a lo largo del día, de acuerdo con los detalles de la construcción, el color y la textura de sus materiales, así como las características de su entorno.

La luz y la forma como se comporta sobre los volúmenes exteriores del edificio, generalmente se considera como parte del diseño arquitectónico, en cuyo caso es un tema que se resuelve desde el programa de construcción. La solución que se adopte, en cuanto al tipo de luminarias y la sombra que se proyecta sobre los muros, debe acentuar la personalidad que identifica a la residencia.

À l'extérieur, la lumière artificielle peut être utilisée pour mettre en relief certaines caractéristiques de la construction comme la couleur, la texture des matériaux employés, le design de la façade, les alentours de la propriété, le portail d'accès, les baies vitrées ou autres éléments jouant un rôle dans l'harmonie et dans l'esthétique de la demeure. La lumière peut être projetée par des lampes placées au pied ou au sommet de la construction, dans les arbres (ou autres supports) ou par des luminaires situés à l'intérieur de la maison.

L'éclairage naturel, quant à lui, illumine l'ensemble de l'espace et on assiste à un jeu entre ombres et lumières tout le long de la journée qui met en valeur la couleur, la texture des matériaux, les finitions et les caractéristiques environnantes de la propriété.

La lumière et la façon qu'elle a de se réfléchir sur la demeure sont généralement considérées comme faisant partie intégrante du design architectural. Aussi doit-on les prendre en compte dès la conception initiale du projet de construction. Les solutions envisagées, le type de luminaires que l'on prévoit, la prise en compte des ombres sur les murs, tout ceci doit servir à faire ressortir le caractère particulier de la demeure.

Im Aussenbereich bietet künstliches Licht die Möglichkeit, bestimmte Details des Gebäudes, Farben, Texturen der Materialien, sowie das Design der Fassade und des Approximationsbereiches, der Eingangstür, Fenster und sonstigen Elemente hervorzuheben, die dem Komplex Harmonie und Schönheit verleihen. Das Licht kann aus Lichtquellen auf dem Boden oder im oberen Bereich des Gebäudes stammen. Eine weitere Möglichkeit ist die Anbringung von Lichtquellen im Inneren des Hauses oder auf Bäumen oder in jeder sonstigen Art von Behälter.

Das natürliche Licht beleuchtet den gesamten Komplex und schafft ein Spiel von Licht und Schatten im Verlauf des Tages, in Übereinstimmung mit den Details des Baus, der Farbe, der Textur und den Oberflächen der Materialien, sowie den Eigenschaften des Umfeldes.

Das Licht und die Form, die es auf die Ebenen im Aussenbereich wirft, wird normalerweise als Teil des architektonischen Designs angesehen. In diesem Fall handelt es sich um ein Thema, das direkt beim Bau mit berücksichtigt wird. Die Lösung, die im Hinblick auf die Beleuchtung und die Schatten gewählt wird, die das Licht auf die Wände projiziert, muss die Persönlichkeit unterstreichen, die das Wohnhaus ausmacht.

White light offers a visual definition of this house's structure and materials.

La luz blanca define visualmente el diseño de la estructura y los materiales de esta casa.

La lumière blanche accentue les effets produits par le design de la structure et les matériaux de cette maison.

Das weisse Licht definiert visuell das Design der Struktur und Materialien dieses Hauses.

TIPS - ASTUCES - TIPPS
• Pools can add a whole new dimension to the front or back of your home, especially if provided with a little light at night.
• Deja espejos de agua para acompañar las fachadas de tu casa y dales un toque de luz en la noche.
• Prévoyez des miroirs d'eau aux pieds des façades. Illuminés la nuit, ils transformeront votre maison.
• Richte Wasserspiegel ein, die die Fassade des Hauses ergänzen und beleuchte sie in der Nacht.

The switch from daylight to artificial lighting transforms the look of the swimming pool and terrace, but at the same time leaves the design intact.

La percepción de la piscina y la terraza se transforman, sin perder la calidad de su diseño, con el cambio de la luz natural por la artificial.

L'aspect de la piscine et de la terrasse, tout en gardant leurs qualités esthétiques, se modifie lorsque l'on passe de la lumière naturelle à l'éclairage artificiel.

Die Wahrnehmung des Schwimmbades und der Terrasse ist durch den Wechsel von natürlichem zu künstlichem Licht geprägt, wobei die Qualität des Designs nicht verloren geht.

TIPS - ASTUCES - TIPPS
• An outdoor fountain with an original design will help you relax with the constant gentle sound of water.
• Escuchar el constante sonido del agua te llevará a la relajación, incluye una fuente original al exterior.
• Le bruit de l'eau d'une fontaine au design original, vous aidera à vous relaxer.
• Das ständige Geräusch des Wassers führt zu Entspannung, beziehe daher einen originellen Brunnen im Aussenbereich mit ein.

TIPS - ASTUCES - TIPPS
- Lighting directed upwards will bathe the wall of the front of your house in light.
- Aprovecha la iluminación ascendente para dotar de barridos de luz a los muros de fachada.
- Des éclairages orientés vers le haut baigneront de lumière les murs des façades.
- Nutze das aufsteigende Licht, um die Mauern der Fassade in Licht zu tauchen.

TIPS - ASTUCES - TIPPS
- *Combining different volumes and lighting can produce a truly spectacular effect at night.*
- *Juega con los volúmenes y la iluminación para conseguir un efecto nocturno conmovedor.*
- *Travailler les jeux de volumes et de lumière permet des effets nocturnes spectaculaires.*
- *Spiele mit den Ebenen und der Beleuchtung, um einen bewegenden Effekt in der Nacht zu erzielen.*

The arrangement of the lighting and the type of light it provides enhance and enrich the façade's volumes, flow and texture.

La disposición de las lámparas y el tipo de luz que emiten, acentúan, para enriquecerlos, los volúmenes, el movimiento y la textura de las fachadas.

La disposition des lampes, la lumière qu'elles projettent, enrichissent les volumes, le mouvement et la texture des murs extérieurs.

Die Anordnung der Leuchten und die Art von Licht, die sie abgeben, heben die Ebenen und die Textur der Fassade hervor und berreichern sie.

Lamps have been placed on the ground to highlight the path leading to the house.

Las lámparas a ras del piso destacan el camino de acceso a la casa, ricamente iluminada.

Les lampes placées au ras du sol bordent le chemin d'accès à la demeure éclairée avec goût.

Die im Boden eingelassenen Lampen erhellen den Zugangsweg zum Haus, der reichlich beleuchtet ist.

TIPS · ASTUCES · TIPPS

- Large glass surfaces can turn your home into a giant lamp.
- Si cuentas con grandes superficies vidriadas convierte tu hogar en una gran lámpara de luz.
- Si votre maison possède de larges surfaces vitrées, celle-ci peut se transformer en une magnifique source de lumière.
- Wenn viele grosse Glasoberflächen vorhanden sind, kann das Haus in eine grosse Lichtquelle verwandelt werden.

HALLS
VESTÍBULOS
ENTRÉES
EINGANGSBEREICH

The hall greets visitors and residents to the house. The way they look offers a taster of the interior decoration, which means all the different elements need to be emphasized with suitable lighting, be it daylight or artificial. If daylight is used, it can be channeled into the inside through windows, skylights, domes or the front door itself. Artificial lighting can be direct, overhead or directed from soffits, embedded lights or tables. Its effect will depend on the type and intensity of the lighting used, as well as the type of items and corners it illuminates. Different types of bulbs will generate different sensations, such as freshness, warmth or comfort.

The best option for halls is to look at how daylight behaves during the course of the day in order to keep the intensity more or less stable by using lights in specific locations or bringing windows or domes into the equation. Artificial light can be used to enhance architectural details, ornamental items and works of art that will bring out the full splendor of the hall.

El vestíbulo da la bienvenida al visitante y a los moradores de la casa. Su apariencia anticipa el diseño interior por lo que es importante que destaquen todos los elementos que lo integran mediante una adecuada iluminación ya sea natural o artificial. Si la iluminación es natural sus fuentes pueden provenir del exterior a través de ventanales, de tragaluces, de domos o de la propia puerta de acceso. La iluminación artificial puede ser directa, cenital o provenir de plafones, lámparas empotradas o de mesa. El efecto que produzca puede variar de acuerdo con el tipo y la potencia de las luminarias y la clase de objetos y rincones que alumbren. Diferentes tipos de focos pueden transmitir sensaciones diversas como frescura, calidez o comodidad.

En estos espacios lo más conveniente es estudiar el comportamiento de la luz natural a lo largo del día a fin de mantener una luminosidad más o menos constante mediante el uso de luminarias colocadas en sitios específicos o la habilitación de ventanas o domos. La luz artificial puede utilizarse para destacar detalles arquitectónicos, elementos decorativos u obras de arte que hacen más atractivo el vestíbulo.

Il est possible de dire que l'entrée est une pièce qui souhaite la bienvenue aux visiteurs et aux occupants de la maison. Son aspect est un avant-goût du design utilisé dans le reste de la demeure. Il est donc important d'en soigner l'éclairage afin que tous ses éléments soient mis en valeur grâce à un éclairage adéquat naturel ou artificiel. Si l'éclairage est naturel, les sources de lumière peuvent se situer à l'extérieur et passer par des baies vitrées, des lucarnes, un toit en verre ou la porte d'entrée. Si l'éclairage est artificiel, il peut être direct, en douche ou émis par des lampes au plafond, encastrées dans les murs ou de table. Les effets produits sont variables car tout dépend de l'intensité utilisée et du type d'objets ou d'endroits éclairés. De même, il existe différents genres d'ampoules qui peuvent donner la sensation de fraîcheur, de luxe ou de confort.

Pour une entrée, il est conseillé de faire attention aux variations de la lumière naturelle tout au long de la journée pour pouvoir ensuite bénéficier d'une illumination plus ou moins constante en utilisant des lampes placées à certains endroits ou en profitant des fenêtres et du toit vitré. L'éclairage artificiel peut aussi être utilisé pour mettre en valeur certains détails architecturaux, certains éléments décoratifs ou une œuvre d'art qui esthétise cette pièce.

Im Eingangsbereich werden die Besucher und Bewohner des Hauses willkommen geheissen. Das Aussehen gibt einen Hinweis auf das Design des Innenbereiches, daher ist es wichtig, alle Elemente, die dort vorhanden sind durch eine geeignete Beleuchtung hervorzuheben, gleichgültig ob es sich dabei um natürliches oder künstliches Licht handelt. Sollte es sich um natürliches Licht handeln, kann die Lichtquelle durch Fenster, Dachluken oder die Eingangstür aus dem Äusseren stammen. Das künstliche Licht kann direkt sein, von der Decke herab leuchten, in die Decke eingelassen sein oder von Tischlampen stammen. Der erzielte Effekt kann unterschiedlich sein, je nach Typ und Potenz der Beleuchtung, sowie in Abhängigkeit von der Art von Gegenständen und den Ecken, die beleuchtet werden. Verschiedene Arten von Glühbirnen können zu unterschiedlichen Eindrücken, wie Frische, Wärme oder Gemütlichkeit führen.

In diesen Bereichen ist es angebracht, das Verhalten des natürlichen Lichtes im Verlauf des Tages zu analysieren, um die Beleuchtung mehr oder weniger einheitlich zu halten, indem Beleuchtung eingesetzt wird, die an bestimmten Stellen angebracht wird oder indem Fenster oder Dachfenster vorgesehen werden. Künstliches Licht kann dazu verwendet werden, architektonische Details, dekorative Elemente oder Kunststücke hervorzuheben, die den Eingangsbereich attraktiver machen.

The visual presence of the lights is minimal, but the lighting they provide plays a vital role in the architectural design and decorative scheme.

La presencia visual de las luminarias es mínima pero la luz que irradian se convierte en un elemento esencial en el diseño arquitectónico y los objetos decorativos.

Les sources de lumiere sont peu visibles mais la lumière qu'elles projettent est telle qu'elle fait partie du design architectural et des éléments décoratifs.

Die visuelle Präsenz der Beleuchtung ist minimal, dennoch verwandelt sich das Licht in ein wesentliches Element des architektonischen Designs und der dekorativen Gegenstände.

TIPS - ASTUCES - TIPPS
- A good combination of shapes and materials can transform your hall into a very appealing area.
- Apuesta a la combinación de formas y materiales en el vestíbulo para dar interés al espacio.
- Une association bien pensée de formes et de matériaux met en valeur l'espace que constitue l'entrée.
- Kombiniere Formen und Materialien im Eingangsbereich, um diesen interessant zu gestalten.

The quality of the light and positioning of the lighting in the soffit bring out the textures and perspective to create a uniform and warm ambience.

La calidad de la luz y la disposición de las luminarias en el plafón, enfatizan las texturas, acentúan la perspectiva al mismo tiempo que crean una atmósfera uniforme y de gran calidez.

La qualité de la lumière ainsi que la disposition des lampions au plafond mettent en évidence les textures tout en accentuant la perspective créant de ce fait une atmosphère uniforme et chaleureuse.

Die Lichtqualität und die Anordnung der Leuchten in der abgehängten Decke, heben die Texturen hervor und betonen die Perspektive, wobei gleichzeitig eine gleichmässige und warme Atmosphäre geschaffen wird.

TIPS · ASTUCES · TIPPS
• You can put sets of objects or modular furniture in your hall to give it an orderly feel.
• Si colocas series de objetos o muebles modulares tu vestíbulo dará la sensación de orden.
• Une lumière de qualité et un éclairage au plafond soulignent les textures et la perspective de la pièce créant ainsi une atmosphère homogène et accueillante.
• Werden Serien von Objekten oder modulare Möbel im Eingangsbereich angebracht, erweckt dies den Eindruck von Ordnung.

The light projected onto surfaces of different materials becomes a key player in the hall's interior decoration.

La luz proyectada sobre superficies de diferentes materiales se convierte en protagonista de la decoración interior.

Une lumière projetée sur des surfaces fabriquées avec divers matériaux est un élément-clé de la décoration intérieure.

Das Licht, das auf die Oberflächen verschiedener Materialien scheint, verwandelt sich in den Hauptdarsteller der Innendekoration.

TIPS - ASTUCES - TIPPS
- *A work of art or a beautiful vase will offer guests a warm welcome.*
- *Una pieza de arte o un buen florero despertarán el sentimiento de bienvenida a tus visitas.*
- *Les visiteurs auront le sentiment d'être bien accueillis si vous placez une œuvre d'art ou un joli vase dans l'entrée.*
- *Ein Kunstwerk oder eine schöne Blumenvase sorgen für einen angemessenen Empfang der Gäste.*

TIPS - ASTUCES - TIPPS
- *Plants and dynamically-shaped objects are a great way to revitalize the entrance hall.*
- *Plantas y elementos de formas dinámicas ofrecerán vitalidad a tu zona de recepción.*
- *Des plantes et des objets aux formes dynamiques donneront de la vie à votre entrée.*
- *Pflanzen und Elemente mit dynamischen Formen verleihen dem Eingangsbereich Vitalität.*

Different atmospheres can be generated by changing the intensity, color or position and angle of the light.

Los cambios en la intensidad, el color o la procedencia de la luz pueden crear atmósferas diferentes.

Les variations de la lumière, ses différentes sources et ses couleurs variées peuvent donner naissance à des atmosphères très différentes.

Die Variationen in Bezug auf Intensität, Farbe oder Herkunft des Lichtes können zu unterschiedlichen Atmosphären führen.

A small table lamp illuminates the wooden panel to unleash the full appeal of the works of art.

Une lampe de table de petite taille associée à une surface en bois fait ressortir l'esthétique des œuvres d'art.

Una pequeña lámpara de mesa proyecta su luz en el panel de madera y destaca el atractivo de las obras de arte.

Eine kleine Tischlampe projiziert ihr Licht auf die Holzwand und hebt die Attraktivität der Kunstwerke hervor.

TIPS - ASTUCES - TIPPS
• A comfortable item of furniture in the hall is a gesture of kindness for your guests.
• Colocar un mueble cómodo en la zona vestibular es un gesto de amabilidad para tus visitas.
• Les visiteurs traduiront la présence d'un meuble confortable dans le vestibule comme une charmante attention à leur égard.
• Ein bequemes Möbel im Eingangsbereich ist eine freundliche Geste für die Gäste.

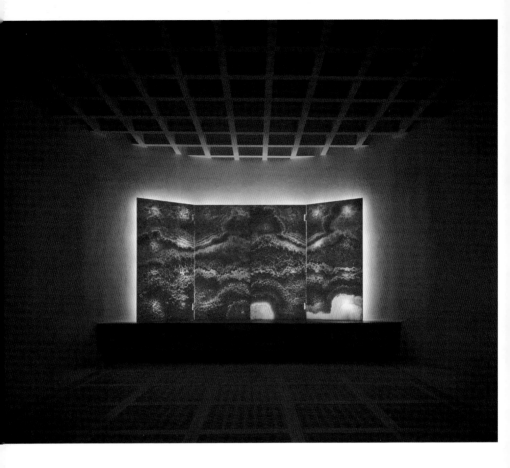

TIPS - ASTUCES - TIPPS
- A hall that offers the unexpected will entice people to enter the house.
- Si tu vestíbulo es sorprendente se antojará siempre el recorrido hacia el interior de la casa.
- On encouragera toujours la découverte du reste de la maison avec une entrée surprenante.
- Wenn der Eingangsbereich originell ist, wird das Interesse dafür geweckt, das restliche Haus kennenzulernen.

LIVING ROOMS
ESTANCIAS
SÈJOUR
WOHNZIMMER

Natural or artificial lighting can generate a unique setting or different ambiences in a single space, by focusing beams onto specific points. This can be achieved with table lamps or foot lamps, or lights embedded into the wall or soffits, that bathe the walls in a flood of light.

The lounge is one of the house's most welcoming and relaxed rooms. At the same time, it is also a public area for friends and relatives to socialize in. The lighting must therefore help create the right atmosphere for each moment. For instance, windows are an excellent option for daylight, which is ideal for a relaxed ambience, while artificial light can be suitably directed and toned to generate a warm and comfortable setting for heart-to-heart chats with close friends.

Low intensity lights buttressed into walls or providing specifically focused light offer a very appealing way to give center stage to paintings or other works of art.

La presencia de la iluminación, natural o artificial, puede crear una atmósfera única o diferentes ambientes en el mismo espacio, mediante haces concentrados en determinados puntos de la habitación, ya sea con el uso de lámparas de mesa o de pie; de luminarias empotradas en el techo o en los plafones que bañan los muros con una cascada de luz.

La estancia es uno de los lugares más acogedores y tranquilos de la casa y al mismo tiempo es un área pública donde conviven la familia y los amigos. La iluminación debe contribuir a crear atmósferas adecuadas para cada momento. Los ventanales, por ejemplo, ofrecen una luz natural de gran calidad que crea un ambiente relajado y distendido, mientras que la luz artificial, convenientemente dirigida y matizada crea ambientes cálidos y confortables ideales.

Las lámparas arbotantes en los muros, igual que las de luz focalizada, de baja intensidad, que suele utilizarse para destacar cuadros u obras de arte, son fuentes lumínicas muy atractivas.

Une pièce peut être caractérisée par une ou plusieurs atmosphères particulières grâce à un éclairage naturel ou artificiel précis. Pour y parvenir, on aura recours à une lumière ponctuelle pour certains endroits de la pièce avec des lampes de table ou sur pied. On peut aussi employer des lampes encastrées dans le plafond pour un éclairage des murs en douche.

Le séjour est une des pièces les plus confortables et paisibles de la maison. C'est aussi une pièce commune où l'on partage certains moments en famille ou entre amis. L'éclairage doit être un des éléments qui contribue à créer une atmosphère particulière pour chaque instant de la journée. Les fenêtres, par exemple, font entrer une lumière naturelle de grande qualité et font de la pièce un endroit propre à la décontraction et à la détente. La lumière artificielle, quant à elle, si elle est bien dirigée et utilisée avec mesure, transforme le séjour en un lieu accueillant et confortable propice à la conversation avec des amis proches.

Enfin, les lampes murales, comme pour un éclairage ponctuel de faible intensité que l'on utilise pour mettre en valeur une œuvre d'art, sont des sources de lumière très esthétiques.

Das Vorhandensein von Beleuchtung, gleichgültig ob natürlicher oder künstlicher Art, kann zu einzigartigen Atmosphären oder zu unterschiedlichen Eindrücken im gleichen Bereich führen. Es können Lichtbündel mit konzentriertem Licht an bestimmten Punkten des Raumes zum Einsatz kommen, gleichgültig ob durch den Gebrauch von Tisch- oder Stehlampen. Leuchten, die in die Decke eingelassen sind, versehen die Wände mit fliessendem Licht.

Das Wohnzimmer ist eines der gemütlichsten und ruhigsten Bereiches Hauses und gleichzeitig handelt es sich auch um einen öffentlichen Ort, an dem die Familie mit Freunden zusammen ist. Die Beleuchtung muss dazu beitragen, eine geeignete Atmosphäre für jeden Moment zu schaffen. Die Fenster stellen zum Beispiel natürliches Licht von grosser Qualität zur Verfügung, das zu einer entspannten und lockeren Atmosphäre führt. Künstliches Licht, das auf geeignete Weise ausgerichtet und abgetönt wurde, führt zu warmen und bequemen Atmosphären, die ideal für ein vertrauliches Gespräch mit den engsten Freunden sind.

In die Wand eingelassene Leuchten sind sehr attraktive Lichtquellen, genau wie gebündeltes Licht mit wenig Intensität, das normalerweise zum Hervorheben von Gemälden und Kunstwerken verwendet wird.

TIPS - ASTUCES - TIPPS
- Bring out the full appeal of your lounge by using differently-shaped furniture and objects in different areas of the room.
- Dale encanto al salón mezclando las formas de muebles y objetos y ubicándolos en distintos ejes espaciales.
- Placer dans plusieurs endroits différents meubles et objets donnera du charme à votre salon.
- Verleihe dem Raum Charme, indem die Formen von Möbeln und Objekten gemischt und auf verschiedenen Achsen des Raumes angebracht werden.

A strip of white light partitions the lounge / dining room in two sections, while carefully directed light highlights the works of art.

Un rai de lumière blanche sépare la pièce en deux alors qu'une lumière bien dirigée fait ressortir la beauté des œuvres d'art.

Una línea de luz blanca divide en dos la estancia comedor mientras la luz convenientemente dirigida destaca las obras de arte.

Eine Linie aus weissem Licht unterteilt das Wohnzimmer und Esszimmer, während geeignet ausgerichtetes Licht die Kunstwerke hervorhebt.

The brightly-lit 3D panel with onyx sheets blends smoothly into the decoration.

El panel tridimensional con planchas de ónix es un espacio luminoso que se incorpora a la decoración.

Une séparation tridimensionnelle lumineuse en onyx fait partie intégrante de la décoration.

Eine dreidimensionale Wand aus Onyx schafft einen beleuchteten Bereich, der sich in die Dekoration einfügt.

TIPS - ASTUCES - TIPPS
• Warm tones will afford intimacy and coziness for your lounge.
• Los tonos cálidos le darán a tu sala un toque de intimidad y la harán acogedora.
• Avec des tons chaleureux, vous ferez de votre salon une pièce intime et accueillante.
• Warme Farbtöne verleihen dem Wohnzimmer einen privaten Touch und machen es gemütlich.

TIPS - ASTUCES - TIPPS
- A harmonious blend of blues and browns in the lounge will strike the right balance between stylish and casual.
- Si incluyes armónicamente azules y marrones en la estancia llegarás al equilibrio entre lo elegante y lo casual.
- Le bleu et le marron, harmonieusement associés, favoriseront une ambiance équilibrée, à la fois raffinée et décontractée.
- Werden auf harmonische Weise Blau- und Brauntöne im Wohnzimmer kombiniert, wird ein Gleichgewicht zwischen Eleganz und Leichtigkeit geschaffen.

Light provides
unification for the
different settings of
this area.

La luz es el
elemento unificador
de este espacio
con diferentes
ambientes.

La lumière
est l'élément
unificateur
des différentes
ambiances créées
dans ce salon.

Das Licht
vereinheitlicht
diesen Bereich, der
unterschiedliche
Atmosphären
aufweist.

TIPS - ASTUCES - TIPPS
- Combining floors is a tricky business, but excellent results can be achieved if done properly.
- La combinación de pisos es arriesgada, pero bien planeada te conducirá a resultados excelsos.
- Associer différents types de sols est risqué mais si vous le faites avec astuce, le résultat sera remarquable.
- Die Kombination von Böden ist risikoreich, aber gut geplant, kann sie zu hervorragenden Ergebnissen führen.

Light, like textures, offers a means of separating the different sections and atmospheres of the lounge.

La lumière, comme les textures utilisées, met en valeur les différentes parties de ce salon avec des atmosphères variées.

La luz, igual que las texturas, propone una diferenciación entre los espacios y distintas atmósferas en la estancia.

Das Licht, genau wie die Texturen, schaffen einen Unterschied zwischen den verschiedenen Bereichen und Atmosphären des Wohnzimmers.

TIPS - ASTUCES - TIPPS
- Daylight and artificial lighting can help you create different ambiences in the same area.
- Las iluminaciones natural y artificial te serán útiles para dotar de distintos ambientes a un mismo espacio.
- Un même espace peut produire des ambiances distinctes grâce à des illuminations naturelle et artificielle.
- Natürliche und künstliche Beleuchtung sind dazu geeignet, einem einzigen Raum verschiedene Atmosphären zu verleihen.

Dynamic lighting draws attention to different points and brings out the full appeal of the room's textures.

Las texturas adquieren mayor atractivo con una iluminación dinámica que pone el acento en diferentes puntos.

L'éclairage dynamique fait ressortir la qualité des textures utilisées en soulignant l'esthétique de certains endroits.

Die Texturen sehen noch attraktiver aus, wenn sie auf dynamische Weise beleuchtet sind, indem verschiedenen Punkte hervorgehoben werden.

The lights play a decorative role and balance the furniture with the overall ambience.

Las lámparas se convierten en objetos decorativos y guardan equilibrio con el mobiliario y todo el ambiente del espacio.

Les lampes se sont transformées en objets décoratifs complémentaires du mobilier et contribuent à l'atmosphère de ce salon.

Die Lampen verwandeln sich in dekorative Objekte und bewahren das Gleichgewicht zwischen den Möbeln und der gesamten Atmosphäre des Bereiches.

TIPS - ASTUCES - TIPPS
• You can experiment with different spatial planes by using short walls and different materials.
• Juega con los diversos planos espaciales valiéndote de muros cortos y cambiando materiales.
• Aménagez différents espaces en utilisant des cloisons de petites tailles et en utilisant des matériaux différents.
• Spiele mit den verschiedenen Ebenen im Raum, und dies unter Nutzung von kurzen Wänden und verschiedenen Materialien.

TIPS - ASTUCES - TIPPS
- *A spiritual atmosphere is ideal for highlighting the harmony of the room.*
- *Un ambiente zen en la sala principal te revelará con rapidez la armonía del espacio.*
- *Une ambiance zen dans le séjour soulignera immédiatement l'harmonie de la pièce.*
- *Eine Zen-Atmosphäre im Wohnzimmer, führt zu Harmonie im Raum.*

Light is focused on specific points of the room to draw the viewer's attention to the wall's texture and the design of the table.

La lumière dirigée avec précision dans cette pièce met en relief la texture des murs ou le design de la table au centre.

La luz se concentra en puntos específicos de la habitación para llamar la atención sobre la textura del muro o el diseño de la mesa de centro.

Das Licht konzentriert sich auf bestimmte Punkte im Raum, um die Textur der Wand oder das Design des Tisches hervorzuheben.

DINING ROOMS

COMEDORES

SALLES À MANGER

ESSZIMMER

Hanging lights whose beams are directed towards the center of the dining room table are conducive to intimacy and enhance the visual appeal of the furniture and decorative objects, especially when bulbs of different types and qualities are combined. Perimeter lighting, when it is the main source of light, usually comes from the soffit to bathe the room in neutral and indirect light. Table lamps and bulbs directed towards specific locations can be used to highlight key points of the room, such as paintings, photographs or other works of art.

The dining room is used at different moments of the day and for different purposes, so the lighting must always be conducive to informal and family socializing or get-togethers with guests. The lighting must help bring out the colors and textures of the food, as well as the dining room decoration. Light from windows or soffits will flood the whole room to generate a relaxed and laidback atmosphere, but if this is insufficient, or as the day goes on, it can be complemented with lights that eliminate or soften the darker zones.

En el comedor, las luces colgantes con haces de luz dirigidas al centro de la mesa, sugieren intimidad y aumentan el valor estético de los muebles y objetos decorativos, sobre todo si se combinan diferentes tipos y calidades de bombillos. La iluminación perimetral, como fuente primaria, que generalmente proviene del contorno del plafón, baña de luz indirecta y neutral la habitación y si se pretende enfatizar ciertos puntos presentes, como pinturas, cuadros, vitrinas o consolas se pueden utilizar lámparas de mesa o focos de luz dirigida a determinados lugares.

El comedor se utiliza a diferentes horas del día y con diferentes propósitos y siempre requiere una iluminación adecuada, ya se trate de una convivencia informal y familiar o una reunión con invitados. La luz debe permitir apreciar los colores y la textura de los alimentos, pero también la forma como está decorado el espacio. La que proviene de ventanas o plafones baña toda la habitación y crea un ambiente relajado y tranquilo; cuando no es suficiente, o conforme avance el día, se complementa con luminarias que eliminan o suavizan las zonas de penumbra.

Dans une salle à manger, des lampes suspendues avec un faisceau concentré sur le centre de la table, font de la pièce un endroit intime et soulignent l'esthétique des meubles et des objets décoratifs, en particulier lorsqu'on utilise différents types de séparations. Un éclairage périphérique, en tant que source primaire de lumière, suit généralement les contours du plafond pour illuminer de manière indirecte et neutre la pièce. Si l'on désire mettre en valeur certains objets de la salle comme des peintures, des tableaux, des étagères vitrées ou des commodes, il est conseillé d'utiliser des lampes de table ou des ampoules dirigées vers des endroits précis. La salle à manger est utilisée à différents moments de la journée pour diverses raisons. Un éclairage particulier est donc nécessaire pour une réunion familiale en toute décontraction ou pour passer un moment avec des amis. L'éclairage doit mettre en valeur les couleurs et la composition des plats servis à table mais aussi la décoration de la pièce. La lumière provenant des fenêtres ou du plafond illumine toute la pièce pour en faire un endroit calme et paisible. Lorsque la pièce s'assombrit parce que la lumière naturelle décline, l'illumination peut être assurée par des lampes qui éclairent complètement ou partiellement les endroits sombres.

Im Esszimmer sorgen Hängelampen mit Lichtstrahlen, die auf die Mitte des Tisches gerichtet sind, für Privatsphäre und erhöhen so den ästhetischen Wert der Möbel und der dekorativen Gegenstände, vor allem, wenn verschiedene Typen und Qualitäten von Glühlampen verwendet werden. Die perimetrische Beleuchtung als hauptsächliche Lichtquelle, die in der Regel an den Rändern einer abgehängten Decke angebracht ist, hüllt den Raum in indirektes und neutrales Licht. Und wenn dann noch bestimmte Punkte im Raum hervorgehoben werden sollen, wie Gemälde, Bilder, Vitrinen oder Konsolen, können Tischlampen oder Leuchten verwendet werden, deren Licht auf bestimmte Stellen gerichtet wird. Das Esszimmer wird zu unterschiedlichen Tageszeiten und für verschiedene Zwecke verwendet und stets ist dabei eine geeignete Beleuchtung erforderlich, gleichgültig ob für ein informelles und familiäres Zusammensein oder ein Treffen mit Gästen. Das Licht muss die Wahrnehmung der Farben und Texturen der Lebensmittel ermöglichen, aber auch der Art und Weise mit der der Raum dekoriert ist. Licht aus abgehängten Decken oder Fenstern erhellt den gesamten Raum und schafft eine entspannte und ruhige Atmosphäre. Wenn dies generell oder im Verlauf des Tages nicht ausreichend ist, kann die Beleuchtung durch Lampen ergänzt werden, die dunklere Bereiche beseitigen oder abmildern.

TIPS - ASTUCES - TIPPS
- Merge your dining room and lounge by using uniform floors and walls, as well as neutral colors for the furniture.
- Integra tu sala-comedor homogenizando pisos y muros y utilizando colores neutros en muebles.
- Harmonisez les sols et les murs, utilisez des couleurs neutres pour les meubles et votre salle à manger sera intégrée à votre salo
- Vereine Esszimmer und Wohnzimmer, indem die Böden und Wände gleich gestaltet werden; die Möbel sollten neutrale
  Farben aufweisen.

The source of light could be an ornamental item, as is the case of this lamp with its sleek, original design and translucent lampshades.

La fuente de luz puede provenir de algún elemento decorativo, como en el caso de esta lámpara cuyo diseño es tan atractivo y original, con pantallas translúcidas en forma de caja.

L'éclairage peut parfois prendre l'aspect d'un objet décoratif comme c'est le cas ici avec cette lampe au design si attrayant et si original à base de panneaux translucides en forme de boîtes.

Eine Lichtquelle kann ein dekoratives Element sein, wie in diesem Fall eine Lampe, deren Design in Form eines Gehäuses aus durchscheinendem Material attraktiv und originell ist.

A classical
style chandelier
affords a touch
of originality and
contrasts with the
straight and simple
lines of the dark
wooden furniture.

Un candil de
estilo clásico
da un toque de
originalidad y
contrasta con
las líneas rectas
y simples de
los muebles de
madera oscura.

Un lustre de style
très classique
apporte une
touche d'originalité
à la pièce et
contraste fortement
avec les lignes
droites et simples
des meubles en
bois foncé.

Ein klassischer
Kronleuchter
verleiht einen
Touch von
Originalität und
kontrastiert mit
den geraden und
einfachen Linien
der Möbel aus
dunklem Holz.

TIPS - ASTUCES - TIPPS
• Spectacular chandeliers are a classical option and highly practical for the dining room.
• Considera la opción de lámparas de araña, siguen siendo objetos clásicos y funcionales para comedores.
• Pensez à un lustre, objet à la fois classique et fonctionnel dans une salle à manger.
• Ziehe die Möglichkeit eines Kronleuchters in Betracht, da dies klassisch und funktionell im Bereich des Esszimmers ist.

The translucent panel in this dining room performs a double lead role as both a rich source of lighting and a wall separating the dining room from the rest of the house.

El panel translúcido asume un doble papel protagónico en el escenario de este comedor, como una rica fuente de iluminación y como un muro que separa el comedor del resto de la casa.

Le panneau translucide joue un double rôle dans cette salle à manger, lui procurant une superbe source de lumière et séparant la salle à manger du reste de la maison.

Die durchscheinende Wand spielt eine doppelte Hauptrolle in diesem Esszimmer, als Lichtquelle und als Mauer, die das Esszimmer vom Rest des Hauses abtrennt.

TIPS - ASTUCES - TIPPS
• A partition can take on a highly decorative role in an area.
• Considera que un muro divisorio puede convertirse en objeto de ornato de un espacio.
• Un mur de séparation peut aussi devenir un objet décoratif dans une pièce.
• Ziehe in Betracht, dass die Trennwand sich in ein Dekorationsobjekt des Raumes verwandeln kann.

This translucent panel acts as a primary source of light, while daylight is toned by the lattice.

El panel translúcido se convierte en una fuente de luz primaria, mientras que la celosía matiza la luminosidad natural.

Une panneau translucide se transforme en source primaire de lumière et les jalousies nuancent la luminosité naturelle.

Eine Wand aus durchscheinendem Material verwandelt sich in die hauptsächliche Lichtquelle, wobei die Jalousie das natürliche Licht abtönt.

Glass is a natural reflector of light and accentuates the luminous impact on the whole area.

El cristal refleja la luz de manera natural y multiplica el efecto luminoso que inunda todo el espacio.

Le verre reflète naturellement la lumière et accentue l'éclairage de toute la pièce.

Das Glas spiegelt das Licht auf natürliche Weise wider und vervielfacht den Leuchteffekt, der den gesamten Bereich erhellt.

TIPS - ASTUCES - TIPPS
• Lounge and dining room furniture should be comfortable and in proportion to the size of the room.
• Intenta que los muebles de sala y comedor sean cómodos y proporcionados al área.
• Pour la salle à manger et le salon, essayez de choisir des meubles confortables et proportionnés à l'ensemble du volume.
• Die Möbel im Wohnzimmer und Esszimmer sollten bequem sein und in Bezug auf die Grösse in den Raum passen.

Focused lighting brightens up small surfaces to create a sensation of warmth and intimacy.

La luz concentrada ilumina pequeñas superficies y provoca una sensación de calidez e intimidad.

La lumière dirigée avec soin éclaire des petites surfaces et font de la salle une pièce intime de grande valeur.

Das konzentrierte Licht erhellt kleine Oberflächen und erweckt den Eindruck von Wärme und Privatsphäre.

TIPS - ASTUCES - TIPPS
• The blend of light and shade, crowned by exactly the right lighting above the table, will create an unforgettable effect.
• La mixtura de luces y sombras, y la iluminación precisa sobre la mesa, emitirán una impresión única.
• Le fait d'éclairer avec soin la table ainsi que les jeux d'ombre et de lumière produisent des effets remarquables.
• Die Mischung aus Licht und Schatten, sowie die Beleuchtung genau über dem Tisch, führen zu einem einzigartigen Eindruck.

CORRIDORS AND STAIRS
PASILLOS Y ESCALERAS
COULOIRS ET ESCALIERS
FLURE UND TREPPEN

Corridors and stairs provide access between different parts of the inside of the house, acting as links or transition areas that join different settings. The fact that little time is usually spent in them does not mean they should be dull or devoid of personality. In fact, they should be designed in a way that offers a taster of the architectural features that characterize the building as a whole.

Because of their role, they need to be kept free of obstacles and well lit, preferably with daylight. If this is not possible, then lights can be positioned all along the floors, walls or ceiling. Decoration should therefore be kept down to a minimum with the visual appeal being provided by the design or textures and often, as is the case of the stairs, simply by the way they look. They can also be a central feature of the inside of the house. These areas should be able to make the most of the daylight entering through skylights or windows, and nearly always use artificial lighting on walls to highlight steps, enrich textures, distinguish decorative items or light up the whole space. The sources of lighting can themselves be objects of great beauty and splendor.

Los pasillos y escaleras son zonas de tránsito al interior de la casa, puntos de enlace o transición entre diferentes ambientes; se permanece poco tiempo en ellos pero eso no significa que deban ser aburridos o carentes de interés. Al contrario, su diseño suele ser una anticipación de la forma que se ha adoptado para resolver arquitectónicamente toda la residencia.

Por su naturaleza deben encontrarse libres de cualquier tipo de obstáculo y bien iluminados, de preferencia con luz natural, de no ser así conviene colocar luminarias a lo largo de su circulación, en los pisos, los muros o los techos. La decoración que pueden contener, por tanto, es mínima, de manera que generalmente basan su atractivo en el diseño o las texturas y en muchas ocasiones, como en el caso de las escaleras, en su aspecto, que puede ser el punto nodal del interior de una casa. Estas áreas pueden aprovechar la luz natural que proviene de tragaluces o domos, a través de ventanales y casi siempre con luminarias artificiales que bañen los muros, acentúen los peldaños, enriquezcan las texturas, destaquen los elementos decorativos o que iluminen todo el conjunto. Las fuentes de iluminación suelen convertirse en un elemento de gran atractivo y belleza.

Les couloirs et les escaliers sont des lieux de passage à l'intérieur d'une maison. Ce sont aussi des points de rencontre ou des espaces de transition entre différentes pièces. On y reste peu de temps mais cela ne veut pas dire que l'on doit en faire des endroits mornes et sans intérêt. Bien au contraire, ces lieux doivent être vus comme les reflets des solutions architecturales adoptées pour le reste de la demeure.

De par leur fonction, ces endroits doivent être vides de tout obstacle et bien éclairés par la lumière naturelle de préférence. Si tel n'est pas le cas, il est recommandé de placer des lampes en suivant leur trajectoire, au sol, sur les murs ou au plafond. Étant donné que la décoration est minimale, les qualités esthétiques de ces espaces de transition résident en général dans la texture des matériaux ou, pour les escaliers, dans la forme, car ces derniers constituent parfois le cœur de la maison. Dans ces lieux de passage, on peut profiter de la lumière naturelle grâce à des lucarnes, des toits vitrés ou des fenêtres. La lumière artificielle, obligatoire pour bien éclairer l'espace entier, sert aussi à mettre en valeur la texture des murs, la forme des marches ou l'esthétique des éléments décoratifs. Les sources de lumière deviennent ainsi des accessoires esthétiques de grande qualité.

Gänge und Treppen sind Durchgangsbereiche im Inneren des Hauses und Verbindungs- oder Übergangspunkte zwischen verschiedenen Atmosphären. Man verbleibt nur kurze Zeit in ihnen, was aber nicht heisst, dass sie langweilig oder uninteressant sein sollen. Ganz im Gegenteil soll ihr Design einen Vorgeschmack auf die Architektur des gesamten Hauses geben.

Aufgrund ihrer Natur sollten sie keine Hindernisse aufweisen und gut beleuchtet sein, vorzugsweise mit natürlichem Licht. Ist kein natürliches Licht vorhanden, sollten Leuchten entlang des Durchgangsbereiches angebracht werden, dies kann am Boden, an den Wänden oder an der Decke erfolgen. Die Dekoration, die in diesen Bereichen Verwendung finden kann, ist minimal. Daher basiert die Attraktivität meist auf dem Design oder den Texturen und in vielen Fällen, wie zum Beispiel bei Treppen, auf deren Aussehen, das ein Knotepunkt der Innendekoration des Hauses sein kann. In diesen Bereichen kann das natürliche Licht genutzt werden, das aus Dachluken und Öffnungen oder Fenstern stammen kann. Ferner werden meist künstliche Lichtquellen verwendet, die die Wände erhellen und die Stufen hervorheben. Gleichzeitig werden die Texturen und die dekorativen Elemente betont oder es wird der gesamte Bereich einheitlich erhellt. Die Lichtquellen verwandeln sich so meist in ein Element von grosser Attraktivität und Schönheit.

Daylight pours in from two points to light up the whole area and bring out the tones of the decoration and textures.

La luz natural, proveniente de dos puntos, ilumina toda la zona y permite apreciar el tono de la decoración y las texturas.

La lumière naturelle, qui provient de deux sources différentes, permet d'apprécier les teintes dominantes de la décoration et les textures employées.

Das natürliche Licht, das durch zwei Punkte scheint, erhellt den gesamten Bereich und ermöglicht es, die Farben der Dekoration und die Texturen wahrzunehmen.

TIPS · ASTUCES · TIPPS
- *Plants and works of art can be used to bring out the full splendor of a spacious passageway set on different levels.*
- *Saca provecho de las zonas de circulación anchas y en desniveles introduciendo plantas y piezas de arte.*
- *Mettez à profit les lieux de passage larges et hauts de plafond en y plaçant des plantes et des œuvres d'art.*
- *Nutze breite Durchgangsbereiche über mehrere Stockwerke für Pflanzen und Kunstwerke.*

TIPS - ASTUCES - TIPPS

• *Banisters made from sheets of glass should highlight the purity of the design.*
• *Cuida que los barandales formados por placas de vidrio unidas evidencien la pureza del diseño.*
• *Faites en sorte que les gardes-corps, formés par des plaques de verre, soulignent la pureté du design.*
• *Achte darauf, dass die Geländer aus Glasplatten die Reinheit des Designs hervorheben.*

The design of the soffits plunges the textures of the walls and floor into shadow. The design of the stairs projects a sense of size and elegance into the setting.

El diseño de los plafones proyecta su sombra sobre las texturas de muros y pisos. El diseño de la escalera transmite dimensión y elegancia al ambiente.

La forme des plafonds permet un jeu d'ombre et de lumière sur les murs et les sols alors que celle de l'escalier agrandit et esthétise la maison.

Das Design der Decke projiziert Schatten auf die Texturen der Wände und Böden. Das Design der Treppe vermittelt den Eindruck von Geräumigkeit und Eleganz.

The play of light
and shadow
brings out the
beauty of the
combined textures
and shapes of
the corridors and
stairs.

El juego de luces
y sombras añade
volúmen a la
combinación de
texturas y formas
en pasillos y
escaleras.

Un jeu d'ombre
et de lumière
rehausse la beauté
des associations
entre textures
et formes dans
les couloirs et
escaliers.

Das Spiel mit
Licht und Schatten
machen die
Kombination
von Texturen und
Formen in Fluren
und auf Treppen
noch schöner.

TIPS - ASTUCES - TIPPS
• Glass can be used to maximize the depth of a corridor by creating different planes.
• Juega con la profundidad de los pasillos creando diversos planos con vidrio.
• Jouez avec la profondeur des couloirs en y créant différents plans au moyen de panneaux de verre.
• Spiele mit der Tiefe der Flure, indem verschienden Ebenen aus Glas gebildet werden.

The works of art perform a highly decorative role by infusing the ambience with character and personality, while the light accentuates their qualities.

Les œuvres d'art, utilisées aussi comme objets de décoration, singularisent et personnalisent l'ensemble des pièces et la lumière joue ici un rôle décisif en soulignant leurs qualités.

Las obras de arte, convertidas en elementos decorativos, dan carácter y personalidad al ambiente, y en este caso la luz juega un papel decisivo para acentuar sus cualidades.

Kunstwerke, die als dekorative Elemente genutzt werden, verleihen der Atmosphäre Charakter und Persönlichkeit. In diesem Fall spielt das Licht eine entscheidende Rolle und hebt bestimmte Qualitäten hervor.

TIPS - ASTUCES - TIPPS
- *The finishes of the passageways can be turned into the ideal backdrop for a spectacular work of art.*
- *Aprovecha los remates de las áreas de circulación para colocar una obra de arte espectacular.*
- *On profitera de la fin de la zone de circulation pour placer une œuvre spectaculaire.*
- *Nutze die Enden der Durchgangsbereiche zum Aufhängen von spektakulären Kunstwerken.*

Little lighting is required for corridors. Options include hanging lights or niches that spotlight decorative items.

En los pasillos que requieren poca iluminación, se puede recurrir a diferentes opciones de como lámparas suspendidas o nichos con luz que destaquen objetos decorativos.

Pour les couloirs avec un éclairage limité, on peut avoir recours à différentes solutions comme des lampes suspendues ou placées dans des espaces creux pour mettre en valeur les diverses décorations.

In Fluren, in denen wenig Beleuchtung benötigt wird, können verschiedene Möglichkeiten Anwendung finden, wie Hängelampen oder Nischen mit Licht, das dekorative Gegenstände hervorhebt.

TIPS - ASTUCES - TIPPS
- *If the staircase is curved, you can save yourself a lot of trouble by using ductile materials for the handrail.*
- *Si tu escalera es curva no te compliques, utiliza materiales dúctiles para realizar el barandal.*
- *Optez pour la simplicité avec votre escalier en colimaçon et utilisez des matériaux souples pour le garde-corps.*
- *Wenn die Treppe gewunden ist, sollten vorzugsweise biegsame Materialien für das Geländer benutzt werden.*

114

Imagination and creativity are the key for creating objects that will blend in with the indoor architecture as both sources of lighting and works of art.

Con una buena dosis de creatividad se consigue crear objetos que se incorporan a la arquitectura interior como fuentes de iluminación y obras de arte al mismo tiempo.

Si l'on sait faire preuve de créativité, on peut parvenir à concevoir des objets qui trouveront leur place dans la décoration en tant que source de lumière et objets d'art.

Mit ein bisschen Kreativität können Objekte geschaffen werden, die sich in die Innenarchitektur als Lichtquellen einfügen und gleichzeitig Kunstwerke sind.

TIPS - ASTUCES - TIPPS

• *Revitalize the lower section of a raised staircase by placing a stunning work of art alongside it.*
• *Dale movimiento a la zona inferior de una escalera volada y colocada al centro con una instalación artística.*
• *Donnez du mouvement à l'espace situé sous l'escalier central d'une pièce en y plaçant une œuvre d'art.*
• *Verleihe dem unteren Bereich einer frei angebrachten Treppe Bewegung, indem darunter ein Kunstwerk aufgestellt wird.*

BEDROOMS

DORMITORIOS

CHAMBRES

SCHLAFZIMMER

The bedroom is the most intimate and personal part of the house and must therefore be defined by the tastes and need for comfort of its inhabitants. It is not just a room for sleeping in; it is also often the appointed place for reflection, rest or intimate chats.

The ambience must be pleasant all day long. The light in the morning is not the same as in the evening, but it can always be toned with the use of curtains or other elements that regulate its intensity. Artificial lighting can come from the ceiling, small bedside table lamps or the wall, and be directed at specific points for specific purposes. The right types and sources of lighting can give spaces a very singular feel, including loft apartments.

The light in the master bedroom tends to be different from the light in children's or teenagers' rooms. In each case, it must be arranged in a particular manner in terms of intensity or the type of lights and screens to be used.

El dormitorio es el sitio más íntimo y personal de la casa y por lo tanto sólo debe responder al gusto y la comodidad de quien la habita. No se trata nada más de una habitación para dormir, también suele ser el rincón predilecto para la reflexión, el descanso o el lugar reservado para la charla íntima.

El ambiente debe ser grato a toda hora del día; la luz de la mañana es diferente a la del atardecer pero en todos los casos siempre será posible matizarla con cortinas o algunos otros elementos que regulen la intensidad. Con iluminación artificial, la luz puede provenir del techo, de pequeñas lámparas de mesa o de pared dirigidas a puntos y propósitos determinados. Una acertada selección de luminarias y fuentes de luz puede individualizar el espacio incluso en un departamento tipo loft.

La luz de la alcoba principal suele ser diferente a la de los dormitorios de los niños o los jóvenes y en cada caso exige soluciones distintas en lo que se refiere a la intensidad de la luz o el tipo de lámparas y pantallas que deben utilizarse.

La chambre est la pièce la plus intime et la plus personnelle de la maison. C'est pour cette raison qu'elle doit convenir aux goûts des personnes qui l'habitent tout en restant confortable. Il ne s'agit pas seulement d'une pièce où l'on dort. La chambre, c'est aussi un lieu propice aux pensées personnelles, au repos et aux conversations privées.

L'atmosphère doit être agréable à tout moment de la journée, et la lumière, qui change du matin au soir, pourra être modifiée grâce à des rideaux ou autres accessoires qui en réguleront l'intensité. Avec un éclairage artificiel, la lumière peut venir du plafond, de petites lampes de table ou murales dirigées vers des endroits précis pour des fonctions déterminées. Un ensemble de luminaires et de sources de lumières choisi judicieusement permet de singulariser une pièce y compris dans un appartement de type loft.

La lumière de la chambre à coucher des parents peut être différente de celles des enfants (adolescents ou non). Mais dans tous les cas, il faut toujours opter pour une illumination, des lampes (ou autres accessoires d'éclairage) et des abat-jours différents.

Das Schlafzimmer ist der privateste und persönlichste Bereich des Hauses und daher muss dieser Raum dem Geschmack und den Bedürnissen der jeweiligen Bewohner entsprechen. Es geht dabei nicht nur um ein Zimmer zum schlafen, es ist vielmehr auch ein bevorzugtes Eckchen zum Entspannen und Erholen, sowie für ein privates Gespräch.

Die Atmosphäre muss zu jeder Tageszeit angenehm sein. Das Licht am Morgen unterscheidet sich von dem am Nachmittag, aber in jedem Fall ist es immer möglich das Licht mit Vorhängen oder anderen Elementen zur Einstellung der Lichtintensität abzutönen. Bei Verwendung von künstlichem Licht, kann dies von der Decke herabscheinen oder aus kleinen Tisch- oder Wandleuchten stammen, die auf bestimmte Punkte und Ziele gerichtet sind. Eine geeignete Wahl an Leuchten und Lichtquellen kann den Bereich individualisieren, und dies sogar in einer Loftwohnung.

Das Licht des Elternschlafzimmer unterscheidet sich normalerweise von dem der Kinder- oder Jugendschlafzimmer und in jedem Fall sind andere Lösungen in Bezug auf die Lichtintensität oder den Lampentyp gefragt.

Light from the background illuminates the whole setting, while ambient lighting highlights specific areas to create a warm atmosphere.

La luz de techo alumbra todo el ambiente y las lámparas resaltan determinadas zonas y crean una atmósfera de calidez.

La lumière du plafond éclaire toute la pièce et les lampes d'appoint font ressortir certains endroits pour faire de la pièce un endroit élégant.

Die allgemeine Beleuchtung erhellt die gesamte Atmosphäre und die Tischlampen heben bestimmte Bereiche hervor und erwecken den Eindruck von Wärme.

TIPS - ASTUCES - TIPPS
• Beige tones for the bed linen and chocolate brown for the furniture provide a stylish alternative.
• Los tonos beiges en ropa de cama y el café chocolate en muebles evidenciarán tu preferencia por la elegancia.
• Les tons beiges du linge de lit et chocolats des meubles sont une preuve d'élégance.
• Beige Bettwäsche und schokoladenbraune Möbel, zeigen deine Vorliebe für die Eleganz.

Venetian blinds tone the intensity of sunlight, and lamps soften the impact of the more shaded areas.

L'éclairage est ici contrôlé grâce à des stores pour les rayons de soleil et des lampes pour les endroits sombres.

Una cortina tipo persiana regula la intensidad de los rayos solares y las lámparas suavizan las zonas de penumbra.

Ein Vorhang vom Typ Jalousie regelt die Intensität der Sonnenstrahlen und die Lampen erhellen sanft die weniger hellen Bereiche.

TIPS - ASTUCES - TIPPS
- *Polished surfaces and light-colored bed sheets will give the bedroom a tidy look.*
- *Para que tu habitación luzca nítida las grandes superficies deben ser pulidas y claras y los textiles blancos.*
- *De grandes surfaces lisses et du linge de lit blanc donneront à la chambre un aspect net.*
- *Damit der Raum einen klaren Eindruck erweckt, sollten grosse Oberflächen poliert und hell sein und die Textilien weiss.*

Daylight generates a bright and uniform ambience throughout the room, while concentrated lighting puts the focus on specific locations.

La luz natural crea una atmósfera luminosa y uniforme en toda la habitación y las lámparas de luz concentrada particularizan ambientes determinados.

La lumière naturelle harmonise l'ensemble de la chambre et en fait une pièce très claire alors que les lampes placées avec soin donnent naissance à différentes ambiances.

Das natürliche Licht schafft im ganzen Raum eine helle und einheitliche Atmosphäre, wohingegen die Lampen konzentriertes Licht abgeben und so bestimmte Bereiche hervorheben. Gestaltung. Es ist ideal für kleine Räume.

TIPS - ASTUCES - TIPPS

- *Sturdy furniture provides bedrooms with character.*
- *Los muebles macizos y robustos imprimirán carácter a tu dormitorio.*
- *Les meubles en bois massif et robuste donneront du caractère à votre chambre.*
- *Massive und robuste Möbel verleihen dem Schlafzimmer Charakter.*

architectonic   arquitectónicos   architectoniques   architektonische

photographic   fotográficos   photographiques   fotografische